John Kasakoff
Willian Ogawa

Traffic volume study in the central region of Campo Grande

John Kasakoff
Willian Ogawa

Traffic volume study in the central region of Campo Grande

Peak times and critical junctions in a city in full development

ScienciaScripts

Imprint

Cover image: www.ingimage.com

This book is a translation from the original published under ISBN 978-613-9-63300-5.

Publisher:
Sciencia Scripts
is a trademark of
Dodo Books Indian Ocean Ltd. and OmniScriptum S.R.L publishing group

120 High Road, East Finchley, London, N2 9ED, United Kingdom
Str. Armeneasca 28/1, office 1, Chisinau MD-2012, Republic of Moldova, Europe
Printed at: see last page
ISBN: 978-620-8-05240-9

SUMMARY

The aim of this study is to assess the capacity of two important junctions on the urban perimeter of Campo Grande/MS in relation to the volume of vehicles travelling through them: Av. Ernesto Geisel with Av. Afonso Pena (intersection) 1 and Av. Afonso Pena with R. Bahia (intersection 2). The methodology consisted of consulting the Campo Grande/MS Transport and Urban Mobility Master Plan, such as the capacity of the roads under study, which is 3900 vehicles per hour, and the most critical intersections in terms of vehicle volume and road capacity, to characterise local traffic and its aspects, such as road capacity, vehicle classification and road classification in relation to use and occupation, as well as a field survey, which consisted of counting the volume of vehicles passing through each intersection and checking traffic light cycle times. The results were structured in tables and graphs and later compared with data from the Campo Grande/MS Transport and Urban Mobility Master Plan. A large number of cars were observed in relation to other vehicles, reaching 87% of the total per hour, and of the 14 directions of the road analysed, 6 exceeded maximum capacity. It was concluded that some of the intersections under study are operating above their initial projected volume, requiring an update to the Transport and Urban Mobility Master Plan, in order to develop an improvement in the city centre and also consider the use of new technologies, such as intelligent traffic lights, used in large cities.

Keyword: Transport and Urban Mobility Master Plan. Manual Counting. Critical roads.

SUMMARY

CHAPTER 1

INTRODUCTION

According to the National Federation of Motor Vehicle Distribution (FENABRAVE, 2011), vehicle sales grew by 235 per cent between 2003 and 2010. With the increase in vehicles and the disorganised growth of large cities, traffic is becoming increasingly chaotic and its planning more challenging.

The vehicle fleet has grown considerably in Brazilian cities over the last 10 years. The problem is that big cities are growing in such a way that their infrastructure can't keep up with this growth.

The number of cars in Brazil grew by 138.6 per cent, while the population grew by just 12.2 per cent (RUBIM, 2013). The result of this is the overcrowding that the roads are able to withstand, causing congestion and discomfort for drivers, implying more time spent commuting.

If you look at the traffic in Campo Grande/MS at peak times, you get the feeling that it has become heavy and slow, leading you to question whether the capacity of the roads in relation to the number of cars passing through is being exceeded or not.

According to Campo Grande's Transport and Urban Mobility Master Plan (PDTMU) drawn up in 2009, 11 streets were already over capacity and others were close to their total limit.

With the aim of contributing to studies for urban planning in Campo Grande/MS, a field survey was carried out to assess the state of the two junctions in relation to the volume of vehicles, in order to draw a comparison with the PDTMU in force.

CHAPTER 2

OBJECTIVES

2.1 GENERAL OBJECTIVE

To assess the capacity of important junctions on the urban perimeter of Campo Grande/MS in relation to the volume of vehicles travelling through them.

2.2 SPECIFIC OBJECTIVES

- Characterise traffic according to vehicle type and quantity at selected junctions;
- Compare the data collected at the selected junctions with the Campo Grande Transport and Urban Mobility Master Plan;
- Suggest ways of improving traffic at the junctions under study.

CHAPTER 3

LITERATURE REVIEW

3.1 WORLD TRAFFIC

In the 19th century, the majority of freight, public and private transport was carried out by horses. In New York, for example, the fleet numbered close to 200,000 horses, and in 1890 this fleet generated around 2,000 tonnes of waste on the city's streets every 24 hours. In London, on the other hand, according to the **Times of** London newspaper in 1894, if there were no changes, by the mid-1950s all the streets would be buried with 2.5 metres of horse waste (RUBIM, 2013).

To minimise this issue, the 1st International Conference on Urban Planning was held in 1898. The main agenda was to find a solution in the world's major cities to the excessive use of horses as a means of transport at the time. But no definitive solution was found. It was only in the 20th century that the issue was resolved, with the spread of car use (RUBIM, 2013).

According to Rubim (2013, p.55), "the expansion of the car has brought countless benefits to the **technological and economic development of cities [...]",** even with its positive and negative points, which are still felt today, it has meant that investment in the sector has remained constant, leading to growing expansion.

A hundred years after the crisis experienced by horses as a means of transport, the use of cars (once a solution) has become a serious problem in large urban centres. And so the problems generated by this evolution, their social, environmental and health impacts, are once again being discussed. The car has become the horse of the 21st century (RUBIM, 2013).

Owning a car, however, is only the tip of the mobility problem in big cities. In a study carried out in 2017, with data obtained in 2016, by the company **TomTom Traffic Index**, there is only one Brazilian city in the ranking of cities with the worst congestion in the world. In 2015, three Brazilian cities were listed in the top 10 (Rio de Janeiro - 4th, Salvador - 7th and Recife - 8th), but in 2016 only Rio de Janeiro, ranked 8th, remained in the ranking, as shown in Figure 1.

WORLD RANK	CITY	CONGESTION LEVEL
1	Mexico City	66%
2	Bangkok	61%
3	Jakarta NEW	58%
4	Chongqing	52%
5	Bucharest	50%
6	Istanbul	49%
7	Chengdu	47%
8	Rio de Janeiro	47%
9	Tainan NEW	46%
10	Beijing	46%

Figure 1 - The world's 10 most congested cities
Source: TomTom Traffic Index, 2016.

According to Figure 1, Mexico City has the highest level of congestion at 66 per cent, followed by Bangkok and Jakarta. Figure 1 shows that the level of congestion is mainly a problem in third world countries, often due to technological limitations.

3. 2BRAZILIAN TRANSIT

National traffic is governed by Federal Law No. 9.503 of 23 September 1997, which establishes the Brazilian Traffic Code. According to Art. 1, all traffic of any kind on the land routes of the territory is governed by this law.

is governed by this code. The objectives of the National Traffic System are to establish guidelines for the National Traffic Policy, standardise technical criteria and establish a system for the permanent flow of information between its various bodies, all of which are set out in Article 6.

In 2001, Federal Law No. 10.257 of 10 July 2001 regulated articles 182 and 183 of the Federal Constitution, established general guidelines for urban policy and made other provisions. This law, known as the City Statute, in its 3rd article establishes the Union's competence to establish guidelines for urban development [...] transport and urban mobility.

Article 4 of Federal Law 10.257/2001 states that the Master Plan is one of the municipal planning instruments for urban policy in Brazil. It would be ideal for all municipalities to draw up a Master Plan, regardless of their number of inhabitants, with a view to planned and structured development, so that citizens can enjoy greater well-being (OCTÁVIO, 2015).

The main guidelines listed by the Ministry of Cities (2004) for an urban mobility policy are:

- Reduce the number of motorised journeys:

- Rethinking urban design;
- Rethink vehicle circulation;
- Develop non-motorised means of transport;
- Recognise the importance of pedestrian movement;
- Provide mobility for people with disabilities and mobility restrictions;
- Prioritise public transport;
- Consider waterway transport;
- Structuring local management.

The Master Plan becomes a Municipal Law, that is, a pact between the population and the legislative and executive powers. Since compliance with legislation is not always carried out in a way that pleases the entire population, this is always a challenge for the three powers that be. That's why the participation of the population is important, especially as a monitoring agent (OCTÁVIO, 2015).

According to data from the Brazilian Institute of Geography and Statistics (IBGE, 2014), only 55.3 per cent of municipalities with more than 500,000 inhabitants have a Transport Plan. Of the remaining 44.7 per cent, which should also have had a plan more than a decade ago, only 28.9 per cent are drawing it up. If the cut-off is applied to all 5,564 municipalities in the country, the percentage drops considerably: only 3.8 per cent have this planning instrument.

3.3 TRAFFIC ENGINEERING

According to (Pingnataro, 1973; Silva 2008), Traffic Engineering involves aspects such as:

- Geometric design;
- Planning;
- Road traffic operation (rural or urban);
- Road network;
- Terminals;
- Interaction with adjacent land use (land use) e;
- Integration between different modes of transport

Vieira (1999) states that traffic engineering aims to provide safe, efficient and convenient movement of people and goods. A more modern concept for traffic engineering would be **"Mobility Engineering", which is based on three factors:** Vehicle Engineering; Road Engineering and Human Factors. According to Vieira (2012, pg. 8).

> Traffic # Transit. TRANSIT - is the movement in general of people and/or vehicles and TRAFFIC, comprises the definition of roads and the movement of people, goods or vehicles through it, by appropriate means in an orderly manner.

Although there is no university degree course in Traffic Engineering, some Civil Engineering courses have this subject on their curriculum. There are also postgraduate and master's programmes in the area, aimed at engineers from any field, architects and technologists (VIEIRA, 2012).

3.4 ROAD CLASSIFICATION

In 1996 there was a decree from the Federal Judicial System establishing a hierarchical road system, separated into arterial, main, collector and local roads, with a view to building cycle paths on the busiest roads. This hierarchisation was altered in 2008 and became part of the Campo Grande/MS Land Use and Occupation Planning Law as an annex.

The terminology was adapted to the Brazilian Traffic Code (Federal Law No. 9.503/1997) in order to unify the attributions and legality in Brazilian traffic, classified as:

- Rapid Transit Route - VA 1: a route characterised by special accesses with free traffic, no level intersections, no direct access to neighbouring plots and no level pedestrian crossings;
- Arterial road: a road characterised by level intersections, usually controlled by traffic lights, with access to neighbouring plots and secondary and local roads, enabling traffic to pass between areas of the city;
- Collector road - VC: a road designed to collect and distribute traffic that needs to enter or leave expressways or arterial roads, enabling traffic to pass through the city's regions;
- Local Route: a route characterised by level intersections without traffic lights, intended only for local access or restricted areas.

In this way, it can be seen that the current hierarchy established by the current plan concentrates the arterial roads in the central area, bringing the collector roads to the surrounding areas, making a connection between them, and configuring the rapid transit routes as connecting axes between the peripheral areas of the urban perimeter and the areas closer to the centre (CAMPO GRANDE, 2016).

3.5 TRAFFIC GENERATING POINTS (TMPs)

According to the Ibero-American Network for the Study of Traffic Generating Hubs (2005, p.35).

> TMPs are sites or facilities of different natures that have in common the

> development of activities of a size and scale capable of exerting great attraction on the population, producing a significant contingent of journeys, requiring large spaces for parking, loading and unloading, boarding and disembarking, and consequently promoting potential impacts.

According to Fonseca (2010), establishments such as shopping centres, supermarkets, hotels, convention centres, theatres, schools, ports and airports are considered to be TMPs. According to Pinto **et al.** (n.d., p. 2), the negative impacts generated by cars on road traffic are:

- Increase in vehicles travelling on the roads;
- Localised congestion, with a consequent increase in air and noise pollution;
- Interruptions to passing traffic on roads located around the development;
- Increase in accidents.

The study of traffic-generating centres is important for understanding car movements and for justifying certain movements at different times of the day. It is also necessary to consider more points, due to the date of publication of the Campo Grande/MS PDTMU, obviously the emergence of new points is foreseeable, such as Shopping Norte Sul Plaza, Feira Central de Campo Grande, establishments **with "Dr/ve-Thru"** and various other points, taking into account the growth of the city in general.

3.6 POPULATION GROWTH AND VEHICLE FLEET IN CAMPO GRANDE/MS

In the national context, Campo Grande is 17th in population volume among the capitals (CAMPO GRANDE, 2016). The city has grown a lot since then, and in 2016 the municipality had 786,797 thousand inhabitants, with an estimated 863,982 thousand inhabitants, and an urbanisation rate of 98.7%, the highest rate in the state (CAMPO GRANDE, 2016).

Constant population growth, the availability of potentially urbanisable areas and the dynamics of real estate development on the one hand, and the need to regularise developments built in areas located kilometres from the urban network on the other, put pressure on successive extensions to the limits of the urban perimeter, set at 26.223 ha in 1969.223 ha in 1969; with the annexation of the Industrial Centre in 1976, the introduction of areas for the construction of housing estates in 1980 and the annexation of the Moreninha III Housing Estate, the perimeter was set at 28,500 ha in 1982, reaching 35,302 ha,82m^2 in 2005; with the change that took place in November 2012, the intraperimeter area became 35,903.53 ha (CAMPO GRANDE, 2016, p. 67).

Based on the data in Table 1, the growth in Campo Grande's vehicle fleet between 2013 and 2016 is 4.39% per year. The population growth rate was 1.77% between 1991 and 2000 (CAMPO GRANDE, 2015). In other words, vehicle fleet growth is higher than population growth.

Table 1 - Evolution of the Vehicle Fleet in Campo Grande/MS, from 2013 to 2016.

TYPE	2013	2014	2015	2016
Cars	248.372	261.988	273.045	281.083
Lorry	13.862	14.390	14.661	14.898
Truck	40.576	43.567	45.659	47.496
Van	14.654	15.671	16.594	17.384
Minibuses	923	962	994	1.010
Buses	2.191	2.215	2.222	2.233
Motorbike	114.443	120.478	125.279	128.908
Others	14.962	15.805	16.518	17.497
TOTAL	**483.039**	**510.255**	**531.781**	**548.475**

Source: IBGE, 2013, 2014, 2015 and 2016.

Table 2 shows that the trend is towards an increase in the motorisation rate, which by 2030 should be close to one car for every 1.34 inhabitants, which currently stands at 2.04 inhabitants per car. For comparison, Curitiba, which had the highest motorisation rate in the country in 2014, had 1.82 inhabitants per vehicle (CAMPO GRANDE, 2015).

Table 2 - Projected population growth and light vehicle fleet in Campo Grande/MS

Year	2010	2015	2020	2025	2030
Population	760.596	806.702	848.794	887.516	923.366
Fleet	314.938	407.973	501.008	594.043	687.078
Index Projection	2,42	1,98	1,69	1,49	1,34

Source: Campo Grande, 2015.

With the increase in the number of vehicles per inhabitant in the city, the time spent travelling in the urban region tends to increase, as the region is already saturated with roads and there is no possibility of widening them.

3.7 METHOD OF VOLUMETRIC COUNTING OF VEHICLES ON ROADS

According to the DNIT (2006), traffic volume is the number of vehicles that pass through a section of the road, or a particular lane, during a unit of time.

There are two types of counting, manual and mechanical. Each has its pros and cons. The parameters for deciding on the ideal method used in each study depend on the purpose and resources raised for the research. Although with the same purpose, counting vehicles, you can get more data depending on the method used, such as the average speed of vehicles or their classification (GOLDNER, 2013).

Counting vehicles on the roads is necessary for this study, since the main objective is to assess the volume capacity of the city's main junctions. Since the number of cars per inhabitant tends to decrease, i.e. by increasing the number of cars travelling on the road, consequently the volume of the road, the concern arises that these roads may be operating above their maximum capacity.

3.7.1 Manual method

Human material is used to carry out the count and allows classification by size, type, among others. A researcher can count

1.1.1 vehicles per hour, provided that the maximum counting time is between 8 and 10 hours (GOLDNER, 2013).

According to Goldner (2013), observers are changed every 2 or 3 hours for reasons of fatigue. Counting is done at intervals of 5 to 15 minutes and spreadsheets are used with pencil notes or manual counters (which accumulate the number of vehicles over the period).

Its advantages are its accuracy, simplicity, number of picccs of information, great flexibility and speed. Its disadvantage is limited coverage (GOLDNER, 2013).

1.1.2 Mechanical method

There are two types of device: intrusive, which is installed in or attached to the pavement surface, and non-intrusive, which does not modify the structure of the road, but is installed above or on the edges of the traffic lanes. Traffic detectors are used for permanent or mobile installation (GOLDNER, 2013).

According to Goldner (2013), its advantages are its low cost per hour, breadth of coverage and precision. However, their disadvantages are that they don't offer much information and their initial investment is high.

1.1.3 Counting at intersections

Counting at intersections is done manually. The movement entering the intersection is counted by movement (direct or conversion), type of vehicle and must be divided into 15 minutes to determine the variations at peak times (Goldner, 2013).

For signalised intersections, movements do not occur simultaneously but in phases (green light). In this case, the cycle time and the division of green time by phase are noted. And so the time interval must be a multiple of the cycle time (Goldner, 2013).

3.8 LOCAL TRAFFIC

According to Campo Grande (2015, p.33)

> The road structure for motor vehicle circulation is made up of streets and avenues with central medians, in both cases with two or more traffic lanes and most of the time parking on the side of the road. The streets and avenues in the central area of Campo Grande have a concentric radial formation, with roads reaching or tangential to the city centre.

In 2015, the urban area of Campo Grande had a total of

3.9 14 km of paved roads (CAMPO GRANDE, 2015), as shown in Figure 2, with a microdrainage and macrodrainage system. However, there is no standardised asphalt or drainage system that fully covers the city, thus causing flooding or the accumulation of rainwater.

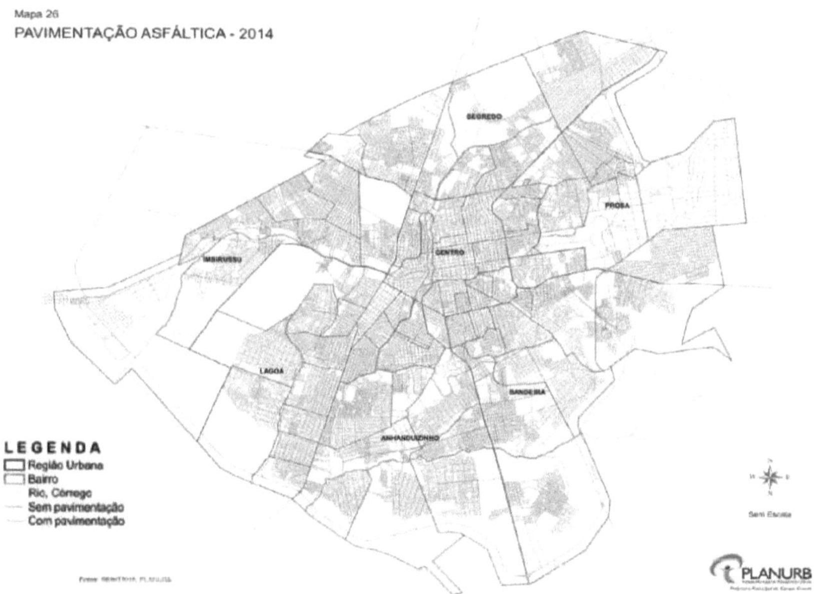

Figure 2 - Paved roads in the urban area of Campo Grande/MS.

Source: Campo Grande, 2016.

> It is common in paved areas for there to be accelerated wear of the pavement, or stretches of road with deteriorated pavement [...] if it is not detrimental to the fluidity and safety of traffic, it makes the implementation and/or maintenance of road signs unfeasible, as well as compromising the comfort of users, especially those of public transport" (Campo Grande, 2015, p.35).

3.9 INTELLIGENT TRAFFIC LIGHTS

In the United States, most traffic corridors operate by combining offline coordination with semi-local action. However, controllers make use of more complex logic geared towards improving the inclusion or exclusion of left-turn stages than the traditional logic available from national controllers (BONETTI, 2001).

Thus, the control may not have a fixed sequence of steps depending on the appearance of requests in the conversions. The usual form of actuation uses dual ring controllers (HCM, 1997), which use actuation in two parallel possible sequences. Another form of control used in the United States (also not available in national controllers) is Volume-Density control (HCM, 1997).

In Australia, before the evolution represented by centralised real-time systems, traffic light controllers were developed for actuated control, related to the search for parameterisations to better meet variations in traffic flows. American Volume-Density control is also used and there is a third actuating technique, called **Waste** Change. In this type of control, the green interruption logic considers the accumulation of excess intervals between vehicles in relation to the basic saturation value, i.e. the average interval in the normal movement of the queue (AKÇELIK, 1995).

According to a study by POLI-USP (2012), the intelligent traffic light is a system for adjusting and synchronising fixed traffic lights, which improves traffic flow by reducing delays and stops by up to 30%. The traffic lights are interconnected by fibre optics, allowing instant communication and adjustment throughout the area. In this way, the equipment adapts to the mobility conditions of each stretch. This is the basic operation, as shown in Figure 3.

1 - The equipment has an on-board controller that analyses traffic conditions in real time and adapts to them

2 - For example, if there is a large flow of vehicles, but the signal is closed and there are no pedestrians to cross, the controller chooses a better time to open that signal

3 - If there are a lot of pedestrians to cross, the traffic light also adapts its opening and closing times to accommodate the entire crossing.

4 - The traffic lights are interconnected and work synchronously.

5 - The computer programme that governs the traffic lights is controlled by a group of operators in a room at Transalvador's Assisted Operation Centre (NDA). They will receive all kinds of information about the traffic lights, such as technical problems with

the equipment, and will speed up the solution.

6 - The equipment makes it possible to adopt plans, such as opening a lane to prioritise the passage of a motorcade or adopting a diversion to avoid congestion caused by an accident.

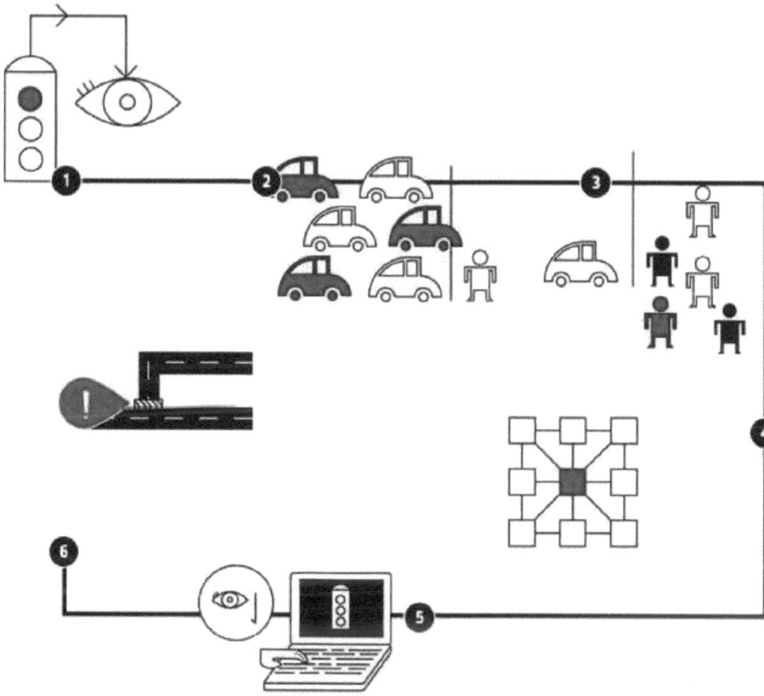

Figure 3 - How smart traffic lights work.

Source: Adapted from Redação, 2017.

According to POLI-USP (2012), the system created a traffic light plan, i.e. it planned the timing of the synchronised signals at six intersections on Avenida Padre Antônio José dos Santos, in the south of São Paulo. With the support of the Traffic Engineering Company (CET), the plan was drawn up for the morning and afternoon rush hours. The improvement in traffic flow was so satisfactory that CET decided to adopt the plans resulting from the Genpolis simulations.

Treating traffic as a living, dynamic organism, the city of Salvador in Bahia began using intelligent traffic lights on 5 May 2017, according to

Redação (2017), with a total of 88 lights spread throughout the city.

The reality in other urban centres seems to be guided by the use of intelligent traffic lights, including Rio de Janeiro, Juiz de Fora and Campo Grande with systems based on detectors and dynamic plan selection, and Porto Alegre, Vitória, Campinas and Ribeirão Preto with systems to supervise traffic light operation, as well as the existing strategy in Curitiba.

CHAPTER 4

METHODOLOGY

According to Campo Grande (2015), the criteria used by Campo Grande's legislators to classify roads was not their function, but their size. This has led to distortions, such as the almost exclusive presence of arterial roads in the city's central area and expanded centre. The current classification of roads should be reviewed, as it should be based on the role that each road plays in urban circulation, as shown in Figure 4.

Figure 4 - Hierarchy of the road system in Campo Grande/MS.

Source: Campo Grande, 2015.

Roads must be designed in relation to the ground. It is unfeasible to develop a speed of up to 60 km/h, in the case of arterial roads, in a place with an excessive number of pedestrians and businesses (CAMPO

GRANDE, 2015).

The classification of roads, despite the flaws in some classifications, leads us to understand the dynamics of traffic in our capital.

According to Campo Grande (2015), the municipality's biggest traffic generating hub is the central region, as this is where the main shopping and service centres are located. This attracts new commercial premises, mainly medium and large enterprises. Four of the capital's main TGPs are Shopping Campo Grande, Colégio Dom Bosco, UNIDERP and the Universal Church of the Kingdom of God.

Figure 5 shows the main generating centres in Campo Grande, including Shopping Iguatemi, which was renamed Shopping Bosque dos Ipês.

Figure 5 - Traffic generating centres in Campo Grande/MS.
Source: Campo Grande, 2015

The junctions selected were Avenida Afonso Pena and Avenida Ernesto Geisel (Junction 1) and the junction between Avenida Afonso Pena and Rua Bahia (Junction 2), as these are the junctions with the highest junction utilisation rates, 86.6% and 84.9% respectively, in relation to road capacity, according to Table 3 (CAMPO GRANDE, 2015).

Table 3 - Volume of vehicles in relation to road capacity

POINT	DESCRIPTION	VOLUME OF VEHICLES IN IN RELATION TO THE CAPACITY OF VIA
31	MATO GROSSO VS CEARA	66,2%
32	MATO GROSSO X 25 DECEMBER	73,1%
33	MATO GROSSO VS RUI BARBOSA	66,8%
34	MATO GROSSO X CALÓGERAS	48,5%

35	MATO GROSSO X ERNESTO GEISEL	53,7%
45	JOAQUIM MURTINHO VS BAHIA	71,5%
46	**AFONSO PENA VS BAHIA**	**84,9%**
47	**AFONSO PENA X ERNESTRO GEISEL**	**86,6%**
48	AFONSO PENA X RUI BARBOSA	63,0%
49	AFONSO PENA X CALÓGERAS	64,6%
50	AFONSO PENA X 14 DE JULHO	60,7%

Source: Campo Grande, 2015.

The municipality of Campo Grande has equipment that enables mechanical quantification, but only in areas where there are speed cameras (also known as pardais). As the stretches under study do not have speed cameras, this method could not be used. For reasons of cost, equipment availability, practicality and time, the decision was made to use manual counting, thus making it easier to characterise the types of vehicles and reducing counting costs.

Traffic characterisation was based on data collected in field surveys. This stage was carried out by collecting data using a spreadsheet containing the types of vehicles, the time of day, cycle time and divided into 15 minutes over 1 hour, which was based on the DNIT (2006) volumetric counting sheet. In this way it was possible to study each traffic movement and count passing vehicles. This gives the variation at peak times, and the value obtained is compared with Table 3.

Data was collected in two periods, during the two traffic peaks, one in the morning from 6.45am to 7.45am and the other in the afternoon from 5.30pm to 6.30pm. The times were chosen on the basis of peak hours (CAMPO GRANDE, 2015) in the morning. In the evening, a time slot was chosen where the authors observed a large flow of vehicles and had the availability for a comparison of the volume of passing vehicles at the chosen times.

Collections were carried out in August and early September 2017, on 14/08, 15/08, 27/08, 30/08, 31/08 and 01/09, with the first 3 days referring to cross 1 and the next 3 to cross 2.

4.1Crossing 1

The roads in question are classified as arterial roads and form part of the city's main avenues. As this is an intersection between two two-way streets, there are 4 directions, separated into A, B, C and D, for better organisation and presentation of the data, as shown in Figure 6. The count was carried out by two people, each responsible for one direction of each avenue.

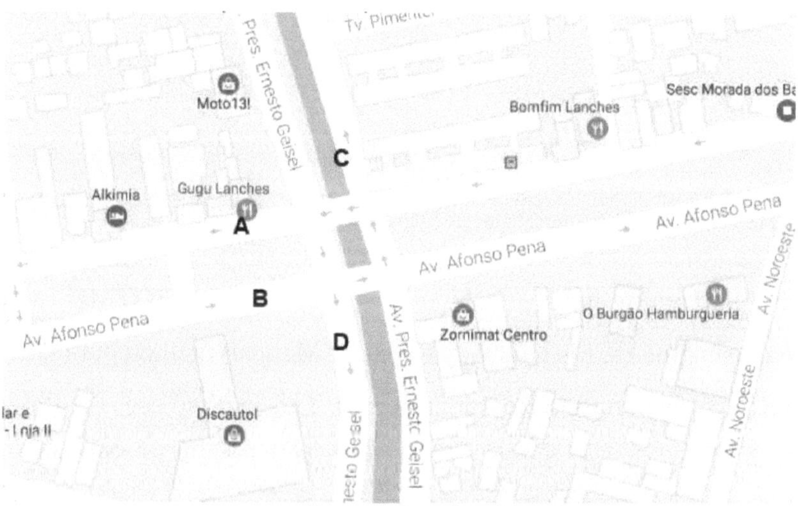

Figure 6 - Location of junction 1

A: Afonso Pena Avenue in the Centre-Barrio direction. B: Afonso Pena Avenue in the Centre-Borough direction. C: Av. Pres. Ernesto Geisel direction Centre-Barrio D: Av. Pres. Ernesto Geisel direction Neighbourhood-Centre. Source: Google Maps, 2017.

4.2Crossing 2

In the case of this junction, it's a little different from the previous junction. It was separated into just 3 directions, A, B and C, because Rua Bahia is one-way, as shown in Figure 7. At this junction, because Rua Bahia is one-way, there was a relay count on this street, and on Av. Afonso Pena it was carried out in the same way as before. It's important to note that at this junction the roads are also classified as arterials, with the same characteristics in terms of the road's function.

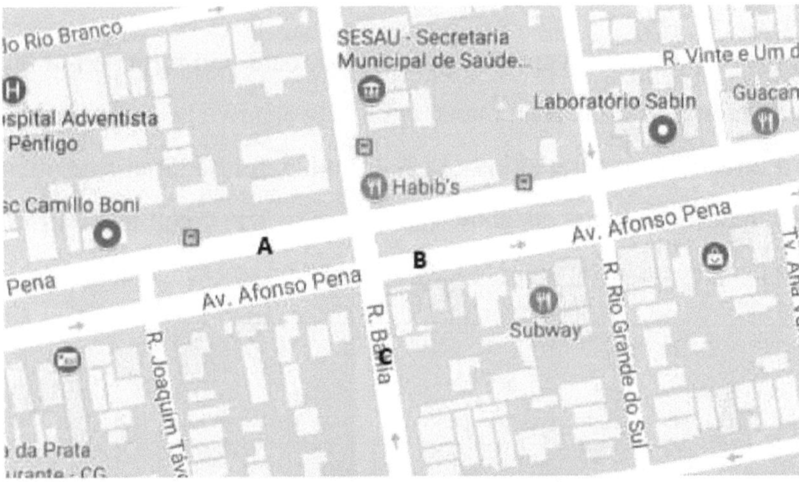

Figure 7 - Location of junction 2

A: Afonso Pena Avenue towards the neighbourhood centre. B: Afonso Pena Avenue towards the neighbourhood centre. C: R. Bahia Source: Google Maps, 2017

The maximum hourly vehicle capacity on the PDTMU was calculated using the HCM (Highway Capacity Manual) method published by the TRB (Transportation Reaserch Board). This method uses the Flow Rate, Peak Hour Factor, Service Flow Rate and Levels of Service to calculate the maximum capacity a road can handle per hour. The roads under study were calculated with a total volume of 3900 vehicles per hour, according to the PDTMU.

After the field research, the data was collated and tabulated using Excel software. The tables were presented showing the volume of vehicles according to their classification and their total and thus compared with the data from the junctions under study.

The total value found on the roads was divided by the maximum volume to obtain the volume of vehicles in relation to the road's capacity. The value found for each lane was used to calculate an average between the junctions, and this value was compared with the values in table 4.

4.3Direction of travel

According to Figure 8, it can be seen that most of the population lives around the central region, which has major traffic generating centres, as mentioned above, a fact that explains the flow of vehicles in different directions depending on the time of day.

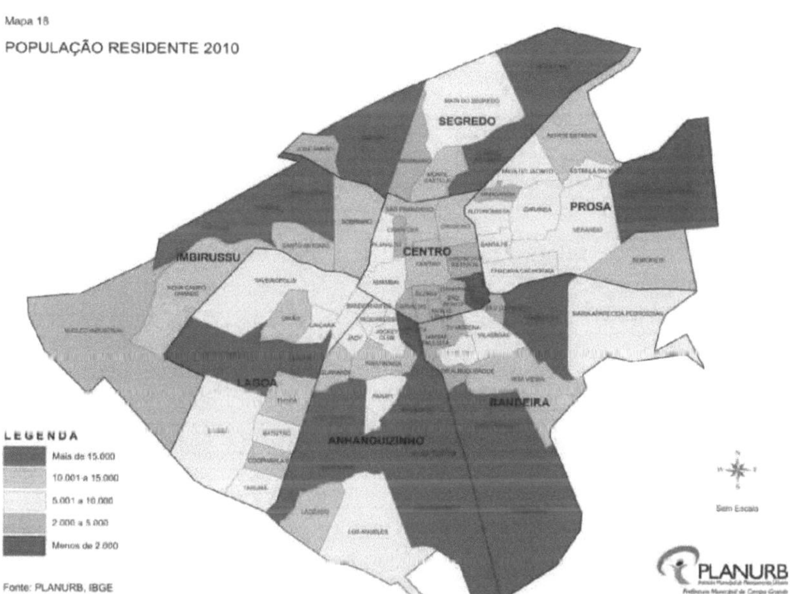

Figure 8 - Resident Population in Campo Grande/MS - 2010
Source: Campo Grande, 2016

CHAPTER 5

RESULTS AND DISCUSSION

In order to analyse the data obtained in the field, tables and graphs were drawn up according to street, direction and collection period.

It is important to note that during the data collection periods, there were no events that could have altered the value found, such as accidents, traffic jams caused by other closed roads, passing lorries or ambulances, or weather events such as rain.

5.1 Crossing 1

Located between the intersection of Avenida Afonso Pena and Avenida Ernesto Geisel, intersection 1 is the most critical, according to the PDTMU. The traffic light cycle on the avenues is 46 seconds on Afonso Pena Avenue and 43 seconds on Ernesto Geisel Avenue.

Figure 9 - Traffic light at Av. Afonso Pena at night time
Source: Authors, 2017

On Afonso Pena Avenue, both in the morning and at night, in both directions, the number of cars was 74.5 per cent on average when compared to the total number of vehicles, as shown in Figure 10.

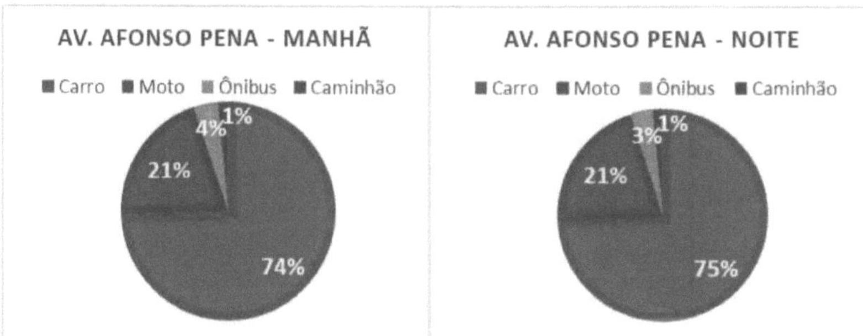

Figure 10 - Percentage according to vehicle classification on Afonso Pena Avenue - Junction 1
Source: Authors, 2017

The same happens on Ernesto Geisel Avenue, with an average rate of 73.5% of cars in relation to the total number of vehicles counted, as shown in Figure 11.

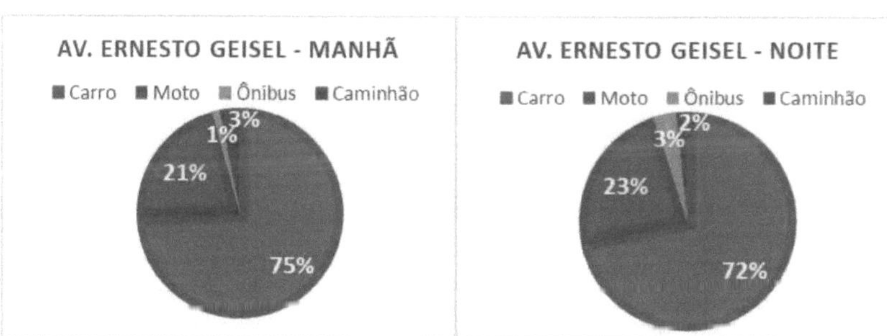

Figure 11 - Percentage according to vehicle classification on Av. Ernesto Geisel - Crossing 1
Source: Authors, 2017

Avenida Afonso Pena is one of the main avenues in the city of Campo Grande, and as such has a large daily traffic flow. Based on the data collected, it can be seen that the neighbourhood-centre commute is greatest in the morning and reverses in the afternoon, since several locations in the central region are considered traffic-generating hubs.

Tables 4 and 5 show the values collected in the field, with the first line (1st) being the values found on the first day of collection, the second line (2nd) the values found on the second day and the third line (3rd) the last day's figure. The fourth line (Average) is the arithmetic mean of the three days of counting, with the total being the sum of all the averages and its % occupancy being in relation to the maximum volume of the road.

Table 4 - Volumetric results for Afonso Pena Avenue and Ernesto Geisel Avenue in the Neighbourhood - Centre direction

	Afonso Pena - *Bairro Centro* - Morning				**Afonso Pena - *Bairro Centro* - Afternoon**			
	Car	**Motorbike**	**Buses**	**Lorry**	**Car**	Motorbike	**Buses**	**Lorry**
1°	3274	1076	143	59	2837	809	157	59
2°	2707	476	130	39	2707	476	130	39
3°	2837	809	157	59	2837	809	157	59
Average	2939	787	143	52	2793	698	148	52
	Total			**3922**	Total			**3691**
	% occupancy			**100,56%**	**% occupancy**			**94,64%**

Source: Authors, 2017.

It was observed that on Afonso Pena Avenue in the morning there was a greater flow of vehicles, caused by the direction of travel from the periphery towards the central region, with a total value of 3922 vehicles per hour, exceeding the total capacity of the road. At the other times, the number of vehicles passing by reduced, but was still alarming, reaching almost 95 per cent of the road's total capacity.

Table 5 - Volumetric results for Av. Afonso Pena and Av. Ernesto Geisel in the Centre - Neighbourhood direction

Afonso Pena - *Neighbourhood Centre* - Morning				**Afonso Pena - *Neighbourhood Centre* - Afternoon**			
Car	**Mot**	**Buses**	**Lorry**	**Car**	Motor	**Buses**	**Lorry**

	orbike			bike			
1° 1787	450	130	59	2850	854	114	78
2° 2707	411	130	39	2922	691	127	20
3° 2837	809	157	59	2837	809	157	59
Average 2443	557	139	52	2870	785	133	52
	Total		**3191**		Total		**3839**
	% occupancy		**81,82%**		**% occupancy**		**98,44%**

Source: Authors, 2017

In the centre-neighbourhood direction, the road's capacity has not been exceeded, but at night it is close to its maximum limit, reaching approximately 99% of the road's total capacity.

As mentioned earlier in the methodology, the volume of 3,900 vehicles per hour was extrapolated, i.e. the road cannot handle the current number of cars that pass through it in the morning in the neighbourhood-centre direction, but even close to the limit, there is still capacity in the other times and directions analysed.

Like Avenida Afonso Pena, Ernesto Geisel plays the same role, but in different directions: while Afonso Pena cuts through the centre in an east-west direction, Geisel cuts in a north-south direction.

Table 6 - Volumetric results for Ernesto Geisel Avenue in the Neighbourhood - Centre direction

Ernesto Geisel - Bairro Centro - **Morning**					**Ernesto Geisel** - ***Bairro Centro*** **- Afternoon**			
	Car	**Motorbike**	**Buses**	**Lorry**	**Car**	Motorbike	**Buses**	**Lorry**
1°	3349	1388	42	136	2819	809	42	136
2°	2819	809	42	136	2986	572	63	94
3°	2986	572	63	94	2986	572	63	94
Average	3051	923	49	122	2930	651	56	108
	Total			**4145**	Total			**3745**
	% occupancy			**106,28%**	**% occupancy**			**96,03%**

Source: Authors, 2017

According to the data collected, in the morning there are 6.28% more vehicles than the road can support, totalling 245 vehicles. This follows the same pattern as in the neighbourhood-to-centre direction, compared to the first period.

Table 7 - Volumetric results for Ernesto Geisel Avenue in the Centre - Neighbourhood direction

Ernesto Geisel - Neighbourhood Centre - **Morning**					**Ernesto Geisel** - *Neighbourhood Centre* - **Afternoon**			
	Car	**Motorbike**	**Buses**	**Lorry**	**Car**	Motorbike	**Buses**	**Lorry**
1°	2351	621	108	84	3202	998	157	73
2°	2895	509	140	42	3014	1109	77	70
3°	3035	865	167	63	3035	865	167	63
Average	2760	665	138	63	3084	991	134	69
	Total			**3627**	Total			**4277**
	% occupancy			**93,00%**	**% occupancy**			**109,67%**

Source: Authors, 2017

Ernesto Geisel Avenue exceeds its capacity during the periods observed, leaving a critical point for traffic in the central region.

In the second period, towards the centre of the city, it's 9.27%, with 377 vehicles overtaken.

Therefore, the total capacity of the junction is greater than that supported by it, and also greater than that reported by the PDTMU, which was 86.6 per cent of the total road capacity.

5.2Crossing 2

The junction between Av. Afonso Pena and Rua Bahia, according to the PDTMU, is the second most critical junction in the city of Campo Grande/MS. As with junction 1, there were no factors that could alter the results collected. The traffic light cycle of each avenue was counted, with Av. Afonso Pena being 50 seconds and Rua Bahia 38 seconds.

Figure 12 - Movement on Bahia Street at night.
Source: Authors, 2017

At the second junction there was an increase of almost 10% in the number of cars compared to other vehicles. The number of buses and lorries decreased dramatically compared to junction 1. At Av. Afonso Pena, an average of 78.5% of cars were analysed compared to the total number of vehicles, as shown in Figure 13.

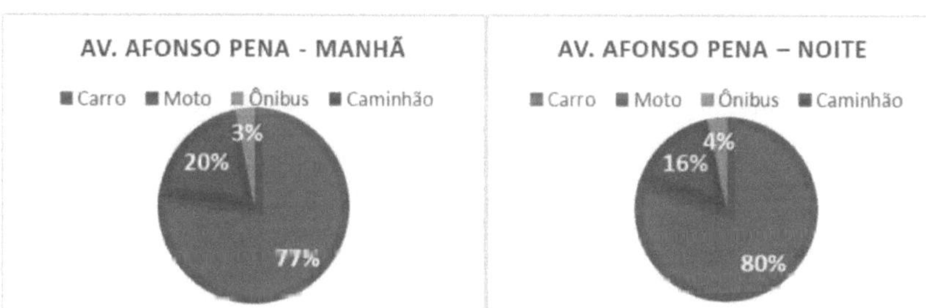

Figure 13 Percentage according to vehicle classification on Afonso Pena Avenue - Junction 2
Source: Authors, 2017

The same happens on Bahia Street, but with an average of 84.5 per cent of cars in relation to the total number of vehicles, as shown in Figure 14.

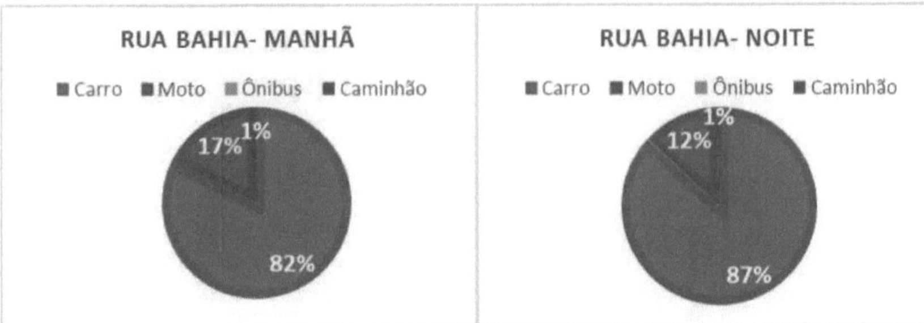

Figure 14 - Percentage according to vehicle classification on Rua Bahia - Crossing 2
Source: Authors, 2017

Despite being the same avenue, this stretch shows a different behaviour to the first, as it is possible to notice that more than a thousand cars pass in one lane, compared to the number that pass in the other lane in the morning. What became clear during the data collection is that at times the avenue is free in one lane and the other side is almost congested.

In both the morning and afternoon periods, there was no higher occupancy rate than that supported by the road, but it is already possible to see a ratio of almost 92 per cent in the second period, as shown in Table 8.

Table 8 - Volumetric results for Av. Afonso Pena and Rua Bahia in the Bairro Centro direction

	Afonso Pena - Bairro Centro - **Morning**				**Afonso Pena** - ***Bairro Centro* - Afternoon**			
	Car	**Motor bike**	**Buses**	**Lorry**	**Car**	**Motor bike**	**Buses**	**Lorry**
1°	2070	306	96	36	2826	378	99	0
2°	2316	318	126	18	3180	570	123	0
3°	2226	408	87	36	2610	**744**	144	18
Average	2204	344	103	30	2872	564	122	6
	Counting:		**2681 vehicles**		**Counting:**		**3564 vehicles**	
	% occupancy		**68,44%**		**% occupancy**		**91,38%**	

Source: Authors, 2017

Table 9 - Volumetric results for Av. Afonso Pena and Rua Bahia in the Centre - Bairro direction

	Afonso Pena - Neighbourhood Centre - **Morning**				**Afonso Pena** - ***Neighbourhood Centre* - Afternoon**			
	Car	**Motor**	**Buses**	**Lorry**	**Car**	Motor	**Buses**	**Lorry**

	Car	Motorbike	Bus	Lorry	Car	Motorbike	Bus	Lorry
		bike				bike		
1°	3096	900	126	0	3276	438	111	18
2°	2946	768	114	18	3522	402	99	36
3°	2970	732	123	0	3372	444	144	54
Average	3004	800	121	6	3390	428	118	36
	Counting:		**3931 vehicles**		**Counting:**		**3972 vehicles**	
	% occupancy		**100,79%**		**% occupancy**		**101,85%**	

Source: Authors, 2017

It can be seen that the maximum volume projected for Afonso Pena Avenue, as well as at the other intersection, is already operating at its limit, creating the hypothesis that perhaps the avenue has several other points where its capacity is compromised. Congestion was visible during the data collection period on the road.

Table 10 - Bahia Street Volumetric Results

	Bahia - Morning				**Bahia - Afternoon**			
	Car	**Motorbike**	**Bus**	**Lorry**	**Car**	**Motorbike**	**Bus**	**Lorry**
1°	3261	711	0	24	2945	466	0	0
2°	3411	568	24	24	3332	411	0	47
3°	3450	703	0	0	3245	521	0	24
Average	3374	661	8	16	3174	466	0	24
	Counting:		**4058 vehicles**		**Counting:**		**3663 vehicles**	
	% occupancy		**104,05%**		**% occupancy**		**93,92%**	

Source: Authors, 2017

Rua Bahia has only one direction and its limit was also exceeded in the morning. It was noted that the number of buses

The number of buses passing along this road is low, reaching approximately 0 on the stretch studied. Few buses were seen in the collection, one of the reasons being that the road is very steep, making it difficult for heavy vehicles to pass.

CHAPTER 6

FINAL CONSIDERATIONS

The capacity of 6 directions analysed are already operating above their maximum vehicle volume capacity and another 6 directions are close to their limit, above 90% of their maximum occupancy, which is 3900 vehicles per hour. However, only 2 movements are operating below 90% of their maximum volume. On Av. Ernesto Geisel there was a percentage of 6.28% more than supported in the neighbourhood-centre direction and 9.67% in the centre-neighbourhood direction.

Because these are the main roads and important intersections in Campo Grande, overloading these roads can cause major inconvenience for the population, from a simple delay to possible accidents caused by heavy and disorganised traffic.

Its value, even if it exceeds that of the road, is still acceptable for a short period. However, like Av. Afonso Pena, it could become a bigger problem in the future. An immediate study and planning at this junction is worthwhile, considering that it is an important junction for the city and that its relief is complicated. The city's PDTMU needs to be updated every 10 years, following the periodicity of the Plan.

As a possible solution, there are intelligent traffic lights, already in use in large cities such as São Paulo. This solution could help improve traffic dynamics by finding the best traffic light cycle times at each junction, thus reducing idle time when travelling. Radars could also be added at strategic points to constantly study the city's main roads, thus helping with future studies.

Based on the studies carried out, as well as the values collected and the comparison with the PDTMU data, it can be seen that the number of vehicles that pass through the junctions analysed on a daily basis is greater than the capacity of the roads there, just as it is necessary to adopt appropriate solutions when the number of cars is absurdly higher when compared to other passing vehicles, solutions that are already available in large cities, such as the use of intelligent traffic lights.

This data provides a basis for further studies in the area, since two points on the same avenue are already operating close to their maximum capacity.

It is therefore necessary to carry out studies into traffic dynamics, such as possible alternative routes, the feasibility of acquiring equipment to distribute traffic generated at peak times, proposing appropriate solutions to improve traffic quality, reducing the time spent travelling and the risk of accidents caused by heavy traffic.

REFERENCES

BONETTI, W. J. **Use and Parameterisation of Traffic-Actuated Traffic Lights.** Polytechnic School of the University of São Paulo, Department of Transport Engineering, São Paulo, Brazil, 2001.

BRAZIL. Law No. 9.503, of 23 September 1997. Establishes the Brazilian Traffic Code. **Federal Official Gazette**, Brasília, 23rd September 1997. Available at: < http://www.planalto.gov.br/ccivil_03/leis/L9503.htm>. Accessed on: 18 Mar. 2017.

____. Ministry of Cities (MCID). **Participatory Masterplan**: a guide for municipalities and citizens. 2004.

____. Law No. 10.257, of 10 June 2001. Establishes General Guidelines for Urban Policy and Makes Other Provisions. **Diário Oficial da União**, Brasília, 10 June 2001 . ____Available at : <http://www.planalto.gov.br/ccivil_03/leis/LEIS_2001/L10257.htm>. Accessed on : 18 Mar. 2017.

CAMPO GRANDE. Municipal Institute of Urban Planning: **Socioeconomic Profile 2016**, Campo Grande, August 2016. Available at: **<http://portal.capital.ms.gov.br/egov/downloadFile.php?id=14382&fileField=arquivo_dow &table=downloads&key=id_dow&sigla_sec=SISGRAN>**. Accessed on 22 March 2017.

. Official Gazette of Campo Grande-MS: **Transport and Urban Mobility Master Plan (PDTMU) of Campo Grande-MS**, Campo Grande, 10 July 2015. Available at: <http://www.capital.ms.gov.br/planurb/downloads/plano-diretor- de-transporte-e-mobilidade-urbana-pdtmu-de-campo-grande-decreto-n-12-681/> Accessed on: 9 May 2017.

DER (2008) **"A Política de Gestão Rodoviária no DER/CE".** Department of Buildings and Roads, Ceará State Government. June/2008 - Document sent to the IDB. Ceará - Brazil.

FONSECA, S. Danilo. **Traffic study at Jackson do Amauri Square**, 2010. State University of Feira Santana, Feira Santana, 2010.

GOLDNER, G. Lenise. **Traffic Engineering**: 1st Module. 2013. Federal University of Santa Catarina, Technology Centre. ECV - 5129 Traffic Engineering. 2013. Available at: <https://ecivilufes.files.wordpress.com/2013/07/engenharia-de-trc3a1fego-mc3b3dulo-1-ufsc.pdf> Accessed on: 21 May 2017.

HCM. TRANSPOTATION RESEARCH BOARD - **Highway Capacity Manual**. National Research Council, Washington, D.C., USA, 1997.

IBGE. Brazilian Institute of Geography and Statistics: **Campo Grande - Fleet,** 2013. Available at: <http://www.cidades.ibge.gov.br/xtras/temas.php?lang=&codmun=500270&idtema=1 28&search=mato-grosso-do-sul|campo-grande|frota-2013> Accessed on: 17 November 2017.

. Brazilian Institute of Geography and Statistics: **Campo Grande - Fleet,** 2014. Available at : <http://www.cidades.ibge.gov.br/xtras/temas.php?lang=&codmun=500270&idtema=1 39&search=mato-grosso-do-sul|campo-grande|frota-2014> Accessed on: 17 November 2017.

. Brazilian Institute of Geography and Statistics: **Campo Grande - Fleet**, 2015. Available at : <http://www.cidades.ibge.gov.br/xtras/temas.php?lang=&codmun=500270&idtema=1 53&search=mato-grosso-do-sul|campo-grande|frota-2015> Accessed on: 17 November 2017.

. Brazilian Institute of Geography and Statistics: **Campo Grande - Fleet,** 2016. Available at : <http://www.cidades.ibge.gov.br/xtras/temas.php?lang=&codmun=500270&idtema=1 63&search=mato-grosso-do-sul|campo-grande|frota-2016> Accessed on: 17 November 2017.

MATO GROSSO DO SUL. Portal of the Municipal Transport and Traffic Agency. **Road system.** Available at <http://agetran.ms.gov.br/agetran/sistema- vi%C3%A1rio>. Accessed on 22 March 2017.

MATO GROSSO DO SUL. Complementary Law no. 76, of 4 November 2005. Regulates the use and occupation of land in the municipality of Campo Grande and makes other provisions. **Hierarchical Circulation Routes**. Campo Grande, 7 November 2005. Available at:

<http://agetran.ms.gov.br/agetran/sites/default/files/anexos/Lei%20Complementar.pdf>.
Accessed on 18 March 2017.

OCTÁVIO, D.P. Thalita. **Urban Policy and Urban Mobility in Brazil**. 2015. Monograph (Bachelor's degree in administration) University of Brasília, Brasília, 2015. Available at:

<http://bdm.unb.br/bitstream/10483/11868/1/2015_ThalitaDalPicoloOctavio.pdf>Accessed on: 29 May 2017.

PIGNATARO, Louis J. **Traffic Engineering: Theory and Practice**. Prentice Hall, Englewood, New Jersey, 1973.

PINTO, André Bresolin; Diógenes, Mara Chagas, Lindau, Luis Antonio. **Quantifying the Impacts of Traffic Generating Hubs**. Article. S.d. Transport Systems Laboratory, Postgraduate Programme in Production Engineering, Federal University of Rio Grande do Sul, UFRGS.

POLI-USP. Polytechnic School of the University of São Paulo: **Semaphore Inteligente,** 27 September 2012. Available at:
<http://www.poli.usp.br/fr/comunicacao/noticias/arquivo-de-noticias/1088-semaforos- inteligentes.html> Accessed on: 27 October 2017.

EDITORIAL. Correio 24 horas: **After test phase, smart traffic lights start working on Monday (8),** 5 May 2017. Available at: <http://www.correio24horas.com.br/noticia/nid/apos-fase-de-testes-semaforos- inteligentes-comecam-a-funcionar-na-segunda-8/> Accessed on: 27 October 2017.

IBERO-AMERICAN NETWORK FOR THE STUDY OF TRAFFIC GENERATING CENTRES (2008). Available at: <http://redpgv.coppe.ufrj.br> Accessed on: 29 May 2017.

RUBIM, Barbara; LEITÃO, Sérgio. **The Urban Mobility Plan and the Future**

of Cities. Estudos Avançados, São Paulo, v.27, n.79, p.55-66, jan. 2013. ISSN 1806-9592. Available at : <http://www.revistas.usp.br/eav/article/view/68702/71282>. Accessed on: 4 June 2017.

SILVA, P. C. da S. **Handout on Elements of Traffic Systems.** UNB, Brasília, 2008.

ANNEX

Annex A - DNIT Volumetric Count Sheet

RODOVIA:BR-000/XX SEGMENTO: km000,0-km000,0

LOCAL	Posto 00 "A"	SENTIDO	Local xxxx - Localyyyy	PESQUISADOR	XXX	
DATA	00/00/2000 2a Feira	PERÍODO	24 HORAS	TEMPO	XXX	FOLHA

		CARROS DE PASSEIO E UTILITÁRIOS		ÔNIBUS		CAMINHÕES			SEMI-REBOQUES					REBOQUES				MOTOS	BICLETAS TRAÇÃO ANIMAL		TOTAL
HORA		P	U	2C	3C	2C	3C	4C	2S1	2S2	2S3	3S2	3S3	2C2	2C3	3C2	3C3	M	B	TA	TOTAL
0:00	1:00																				0
1:00	2:00																				0
2:00	3:00																				0
3:00	4:00																				0
4:00	5:00																				0
5:00	6:00																				0
6:00	7:00																				0
7:00	8:00																				0
8:00	9:00																				0
9:00	10:00																				0
10:00	11:00																				0
11:00	12:00																				0
12:00	13:00																				0
13:00	14:00																				0
14:00	15:00																				0
15:00	16:00																				0
16:00	17:00																				0
17:00	18:00																				0
18:00	19:00																				0
19:00	20:00																				0
20:00	21:00																				0
21:00	22:00																				0
22:00	23:00																				0
23:00	24:00																				0

MIX
Papier aus verantwortungsvollen Quellen
Paper from responsible sources
FSC® C105338

Printed by Books on Demand GmbH, Norderstedt / Germany